HOME
{SWEET}
HOME

ARKANSAS *rescue* DOGS & THEIR *stories*

written by /
GRACE VEST

Little Rock, Arkansas 2017

Copyright © 2017 Grace Vest

All rights reserved. No part of this book may be reproduced without express permission of the publisher, except in the case of reviews. Contact publisher for permission under any circumstances aside from book reviews.

Published in the United States of America by:
Et Alia Press
1819 Shadow Lane
Little Rock, AR 72207
etaliapress.com

ISBN: 978-1-944528-02-7
Library of Congress Control Number: 2017954259

Edited by Erin Wood
Cover & Layout Design by June Upton
Photography by Whitney Bower

{ for jake } - thank you

for loving the dog park

TABLE

OF

CONTENTS

/

25 dogs

/

25 rescues

+
06
———

{ intro }
by grace vest
_____

+
08
———

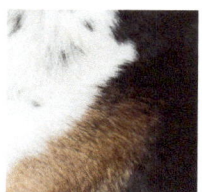

{ part I }
champ, rascal, hattie, kali, walter, & margie
_____

+
22
———

{ part II }
rocky, tripp, lucie, django, piper, & faye
_____

+
36

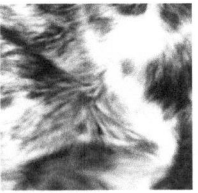

{ part III }
cash, nala, roxy, riley, & frankie

+
46

{ part IV }
maxwell, marshmallow, lyla, & cedar

+
56

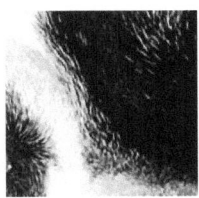

{ part V }
priscilla, sheva, grux, teddy, & georgia

+
68

{ resources }
arkansas animal rescue organizations

+ stephanie parsley photography

{ INTRODUCTION }

Dogs are simple souls. They don't care what you look like, what you do for a living, or how old you are. They're most interested in what's in your heart. Many would agree that if humans acted more like dogs, the world would be a better place. With all that dogs give us, humans should in return give dogs a proper and loving home. Sadly, all dogs aren't treated the way they deserve to be.

Throughout my life, my family has rescued pets. I remember many occasions when my dad would take my sister, my brother, and me to the local Humane Society to pick out our new furry family member. As many who have rescued dogs can attest, canine rescues human as much as human rescues canine. I can say this with authority, because I owe life as I know it to a couple of rescue pups. Rescue dogs led me to my husband, Patrick. If I had never rescued Rascal, I wouldn't have met Patrick and his rescue lab, Jake, at the dog park.

Jake was a protagonist in our fairytale story. He was Patrick's baby; wherever Patrick went, Jake was alongside him. Later, when Patrick and I were dating, Jake became my sidekick. He was my passenger for errands, my beach buddy when Patrick couldn't go to Florida. He loved being outside, but didn't enjoy exercise like I did, so he would stay behind while I ran. Jake was even a best man in our wedding. If it weren't for Jake, I wouldn't have my little family as I know it today. Sadly, he left us in November of 2015.

It's important to me to carry on Jake's memory. I want our son, Townsend, to know how much Jake was loved, and to understand how the simplest things in life—like tail wags, excited sniffs, and head scratches by the front door—can be the best, the things that turn a house into a "home sweet home" for rescued canines and rescued humans.

Jake inspired me to write this book and tell other Arkansas rescue dogs' stories. For the rescue dogs that have become members of my own family, I wanted to write this book. Because of all those I cannot rescue, but that I know are out there grieving for loving homes, I couldn't not write this book. When I started this project, I couldn't have imagined that nearly 300 rescued humans would write in with the stories of Arkansas dogs.

One of those who shared her dog's story was Whitney Bower. A photographer talented in capturing the hearts of her subjects, Whitney's love for dogs and passion for finding homes for rescues made her the ideal choice to be this book's photographer.

It was a challenge to narrow so many powerful stories down to a fraction of those submitted. In spirit, I am including each and every one. I hope that as you read through those I ultimately selected to print, and see their personalities reflected in Whitney's images, you'll think also of the many wagging tails across the state and beyond awaiting loving homes. If someone reads this book and decides to adopt or foster, volunteer at or donate to a local shelter or organization, then I have done my job in sharing Jake's legacy.

———

EVERY RESCUE DOG HAS A STORY. I'M HUMBLED TO SHARE A FEW WITH YOU.

———

/ grace vest

{ PART I }

"I can't imagine how Kali ended up on the path she did, but we are the lucky ones! She exemplifies dignity, and the fact that anyone can overcome hardships!"

/ tracy abston, KALI

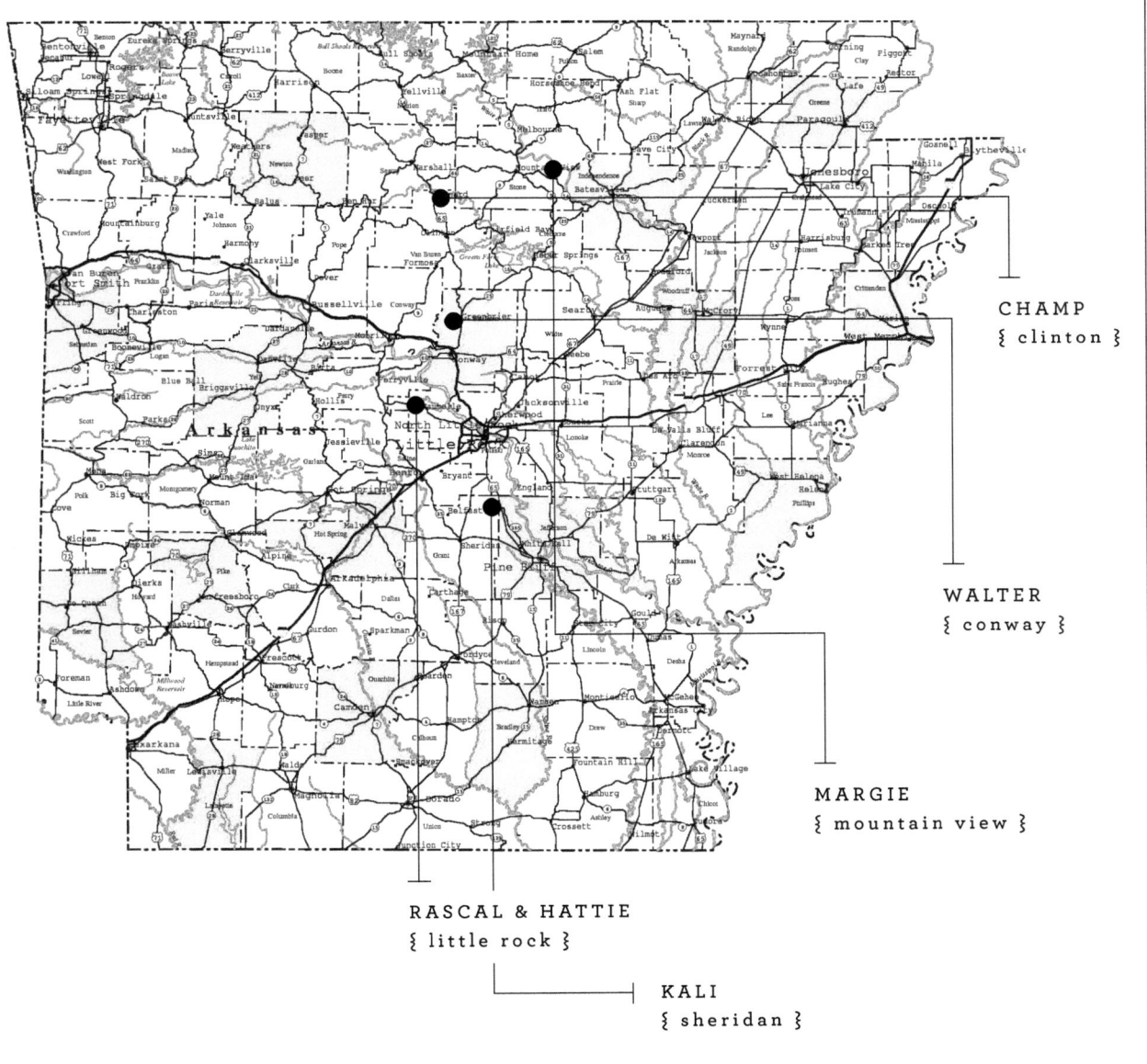

{ CHAMP } | owners / patrick & grace vest

If it weren't for this book, I wouldn't have Champ. My bowlegged little man had a rough start at life. He had a home, but not a good one. Whitney Bower, this book's photographer, called with bad news: a friend of hers had seen a beagle chained to a fence in Clinton, and he'd been there for at least five days. To make matters worse, it looked like the people who lived there had abandoned him after a recent house fire.

After learning about Champ, Whitney stopped what she was doing, drove to Clinton, and rescued him. When she got to the burned-down house, she saw Champ lying down by a joke of a dog house, pieces missing from the roof and nails sticking out from the top and sides. He was so still she thought he was dead. Thankfully, when she approached him he got up. He looked at her as if saying, "Please, take me."

His collar and chain had become embedded in his skin, and he smelled like he had never been bathed. Whitney knocked on the neighbor's door and asked if anyone owned the dog. The neighbors said Champ had belonged to the people that once lived in the burned-down house, and they believed he had been outside on the chain for at least four years. They would give him water and throw him scraps, but said they didn't want to take him for fear of the cops coming to their house.

Whitney loaded Champ in her car, bathed him, took him to the vet, and called me. I took Champ in as a foster dog until we could find him a proper home. As many do, I became a foster failure. Champ loved our rescue lab, Norma, and got along perfectly with our son, Townsend, who was eight months old at the time. It took Champ about three weeks to figure out what love was, because it was clear he had never been loved properly before. We had him fixed and made sure the cut where his collar had once been was healed. He was scared of the couch at first, but now it is one of his favorite spots. Through fostering, we formed a bond that we decided we couldn't break.

Champ is the perfect addition to our family. Whitney remains his forever fairy godmother.

{ RASCAL } | owners / corley & jennifer townsend

My dad would always say, "The best dog is an adopted dog." I was at a point in my life where I was living on my own, and felt prepared to take on the responsibility of having my own dog. I went to the Little Rock Animal Village, adopted a 10-week-old Border Collie mix, and named him Cooper.

Cooper was found in a shed in southwest Little Rock where his brother and sisters had passed away. The pups' mother was nowhere to be found. Cooper was sweet, and we instantly bonded. One night after about a week together, and seemingly out of nowhere, Cooper started foaming at the mouth. He was having trouble breathing. I rushed him to the vet. While we were still in the waiting room, little Cooper passed away in my arms.

Following Cooper's death, it was confirmed that he had suffered from parvo. In puppies, parvo is caused by the canine parvovirus. This virus is highly contagious and spreads through direct contact with an infected dog or by indirect contact with a contaminated object. A puppy is exposed to the parvovirus every time he sniffs, licks, or consumes infected feces. In most cases, young puppies die from this awful virus.

After about a month, I went back to the Little Rock Animal Village, and shared the story about Cooper and his siblings in the shed. They took me to a room and showed me a litter of puppies that were recently born. My sister helped me pick the runt of the group, based on his playful and joyful demeanor. I named him Rascal.

Often from bad comes good, and that was the case with Rascal. He accompanies me everywhere, including to Arkansas Razorback football games. Rascal has been with me through every stage of learning how to "adult." Mentioned in the introduction of the book, Rascal is the reason I met my husband, Patrick, and his rescue dog, Jake, who would become our dog Jake. I took Rascal to the dog park every day, and Patrick brought Jake. We all instantly connected but it wasn't until six years later that Patrick and I finally started dating. During the years between meeting and dating, I called Patrick my "dog park boyfriend" because I could never remember his real name. Our wedding and our son later, we fully appreciate that if it weren't for rescuing Rascal, we may never have met.

These days, Rascal lives with my dad, Corley Townsend. He went to live with my parents when my apartment management decided dogs were no longer allowed. It was intended to be temporary, but Rascal and my dad formed a quick bond. Dad had recently had lost his rescue, Eli, and Rascal was there for him when my dad needed him most. Rascal and I still see each other weekly and Rascal has formed a bond with our son, Townsend. Rascal plays golf at the First Tee of Arkansas, enjoys long weekends at the lake, and loves giving hugs to everyone who'll let him.

{ KALI } | owner / tracy abston

Kali, a Catahoula Mix, was adopted from the Arkansas Paws in Prison Program (PIP) at Tucker Maximum Security Prison in April 2015. Kali was two years old when Tracy Abston met her and immediately fell in love. Kali was rescued by Last Chance Arkansas from a high kill animal shelter in Sheridan.

Kali is large, black, and was very pregnant at the time—all factors making her adoption less likely. Within four hours of her rescue from the shelter, Kali gave birth to three glass-eyed puppies. She was a fantastic mother. All of her puppies have very calm temperaments and are being trained as service or therapy dogs.

After the puppies were weaned, Kali entered the Arkansas Paws in Prison Program. In the middle of the program, one of her teeth had to be pulled by the vet. That's where Tracy met her and immediately fell in love. Tracy always adopted black dogs and she had recently lost her former rescue, Beaux. Tracy knew Kali had been waiting for her family and Kali's new brother Maxwell, who had been found on the side of the road in Arkadelphia with two broken legs. Kali was a blessing for Maxwell because he needed another "dog friend." He became very solitary after Beaux passed and Kali gave him a new lease on life.

Unfortunately, Kali never got to experience an official graduation from Paws in Prison since she was adopted in the middle of it. Her trainer wrote Tracy a wonderful three-page letter all about Kali: what Kali had learned during the program, a list of all her tricks, what she loves, and her story. Her trainer at Tucker Max said, "I never got to say good-bye, but heard Kali went to a great home." He just wished he could have hugged her goodbye. Tracy has written him back with pictures of Kali and Max. She has offered for Last Chance Rescue to take Kali back to Tucker Max to say goodbye if they ever want to.

Kali is absolutely beautiful inside and out, loves everyone, and is sharp and eager to learn.

{ WALTER }

owner / amy turpen

Walter is made of love ... and a lot of hair. He is a beautiful four-year-old great Pyrenees / Labrador retriever mix, adopted by his family through Central Arkansas Rescue for Animals (CARE) in October 2013. He weighs around 90 pounds, but thinks he is tiny. He must be among the sweetest dogs in the world, and is highly intelligent. Very verbal, he talks all the time; sometimes his family even knows what he is telling them. They are sure he always understands what people say to him.

Walter loves to play in the back yard with his sister, Gemma—also a rescue. Great Pyrenees are considered guardian dogs and are bred to watch over sheep. The guardian traits are apparent in him every day. He loves kids and watches over them very closely. The younger they are, the closer he sticks to them. His favorite spot to lie is under the dining room table at home where he can see all three doors, a good vantage point to watch over the house and his people.

Walter was picked up roaming the Conway city streets at about 9 months old, and was a tall and skinny 70 pounds. CARE rescued him from Conway Animal Control and he was accepted into the Arkansas Paws in Prison Program (PIP). Shortly after entering PIP, he was moved to the service dog training program because he was so smart and trainable, and had the perfect temperament. Entering the program meant

he was no longer available for general adoption. But after a few weeks, his trainers noticed he was having some muscle tremors and leg weakness. So, he was pulled from the service dog program and once more made available for general adoption. The trainers continued to teach him as many tricks as they could, and by graduation, Walter knew a long list of tricks.

After vets inspected the tremors and weakness, Walter was diagnosed with an autoimmune disease. It looked like he would be on medication for the rest of his life to keep the symptoms under control. Miraculously, a few months after placement in his forever home, Walter was pronounced 100% healthy.

His short-lived illness remains a mystery, but the consensus is that he had growing pains because he is such a large breed, exacerbated by being so malnourished as a puppy. What we do know without any doubt is that he was meant to be part of Amy Turpen's family. One of the many things they love about him is that he still knows a handful of unusual tricks, ones he learned in service dog training. His favorite is "Dry"; when he's soaking wet, he'll shake dry on command.

{ MARGIE }

owners / lisa & troy schulte

"Everyone should have the good fortune of having one really good dog as a part of their life." Lisa Schulte's father-in-law texted this message to Lisa the day she and her husband, Troy, had to put their first rescue dog to sleep.

Kona was Lisa's first dog to have on her own after college. Lisa's friend Barry had raised Kona from a puppy after she was abandoned in a box at an apartment complex. She was smart, beautiful, and one of the best friends Lisa could ask for. After Kona's cancer diagnosis and seizures at just four years old, Lisa and her husband had to make the difficult decision to put her down. Following her death, Lisa had never been so devastated and her heart remained heavy. Although Kona was irreplaceable, she knew she and Troy needed another dog in their lives.

The couple had heard about Arkansas Paws in Prison, which pairs shelter dogs with inmates in the Arkansas prison system. For 8 weeks, inmates train the dogs on basic obedience commands. The dogs "graduate" at the end of the program, and become available for adoption. Lisa had always admired the program, and loved looking at the pictures of the adorable dogs on the Paws in Prison Facebook page.

A few days after Kona's death, Lisa came across a photo of Margie online. The second Lisa saw the

picture, she knew Margie was going to be their next dog. Lisa commented on the picture, asking if Margie was adoptable, and discovered she was. Lisa promptly arranged a meet-and-greet. Lisa and Troy not only met Margie, they met her inmate trainer. They felt completely at ease during their meeting and were excited to get to know everyone who had worked with their soon-to-be family member.

With some research of the Stone County Humane Society Facebook page, Lisa found that Margie and a dog suspected to be her mother were thrown from a moving car on Christmas Eve in 2015. Margie lived at the Stone County Humane Society from the time she was dumped until she made her way into the Paws in Prison Program in July 2016.

Adopting Margie has been one of the best decisions Lisa and Troy have made. Though she does have a penchant for destroying books and shoes, Margie loves to cuddle and keep them company. She is most likely part pit bull, part Labrador retriever. Don't let the pit bull part scare you, because she is the sweetest dog you will ever meet. Lisa is so glad Margie is a part of their little family.

{ HATTIE } | owners / corley & jennifer townsend

This girl has such a fun and loving soul. You would never guess she was once a stray on the streets. This sweet gal got a second chance at life after my dad decided he needed to have his car washed one Saturday afternoon.

After a Saturday round of golf, my father, Corley Townsend, decided his car was long overdue for a proper wash. When he arrived at the carwash off of Colonel Glenn Road in Little Rock, he noticed a tiny dog hanging out in the back of the building. He asked a few employees if they knew the dog and they said she had been there for about two weeks. They would throw her leftover food and give her water from time to time. They would watch her look both ways, cross the street, and hang out at a local tire place. The employees there would give her more scraps of food, and then she'd return to the carwash.

The carwash was no five-star hotel. The back was nasty, coated in colored soap and dirt, and not suitable living quarters for a tiny dog. Dad could tell she was malnourished, and she was covered in fleas and ticks. He suspected she was pregnant. Deciding that enough was enough, he took her home.

That Monday, he took her to the vet to get shots and a cleaning, discovering she was under a year old and very pregnant. She was only about 13 pounds so it was hard to believe that puppies were living inside of her.

My step-mom, Jennifer, posted the pregnant pup on the Arkansas Lost and Found facebook page, and was contacted by the owners. They couldn't verify any information about the dog and couldn't explain why they couldn't properly take care of her. Dad and Jennifer decided she was not going back to that home, and the previous owners agreed to let them keep her. They named her Hattie, inspired by their recent meal at Hattie B's Hot Chicken in Nashville, Tennessee.

About three weeks later, Hattie had the five beautiful babies pictured. She delivered three at home, but was rushed to the emergency vet in the middle of the night because she was having a hard time birthing the others. Sadly, she lost one puppy, but the others were safely delivered.

All five puppies are larger than Hattie, and have been adopted by loving families. Three of the puppies live in the neighborhood, so mamma and pups occasionally have play dates. Dad and Jennifer cannot imagine life without Hattie, and she is great addition to their now three-dog family!

{ PART II }

"I said, 'Please, God, show me what you want me to do with my life.' Seconds later, I saw a little dog run across the street. I stopped and walked up to her, and she literally jumped into my arms."

/ anna coon, PIPER

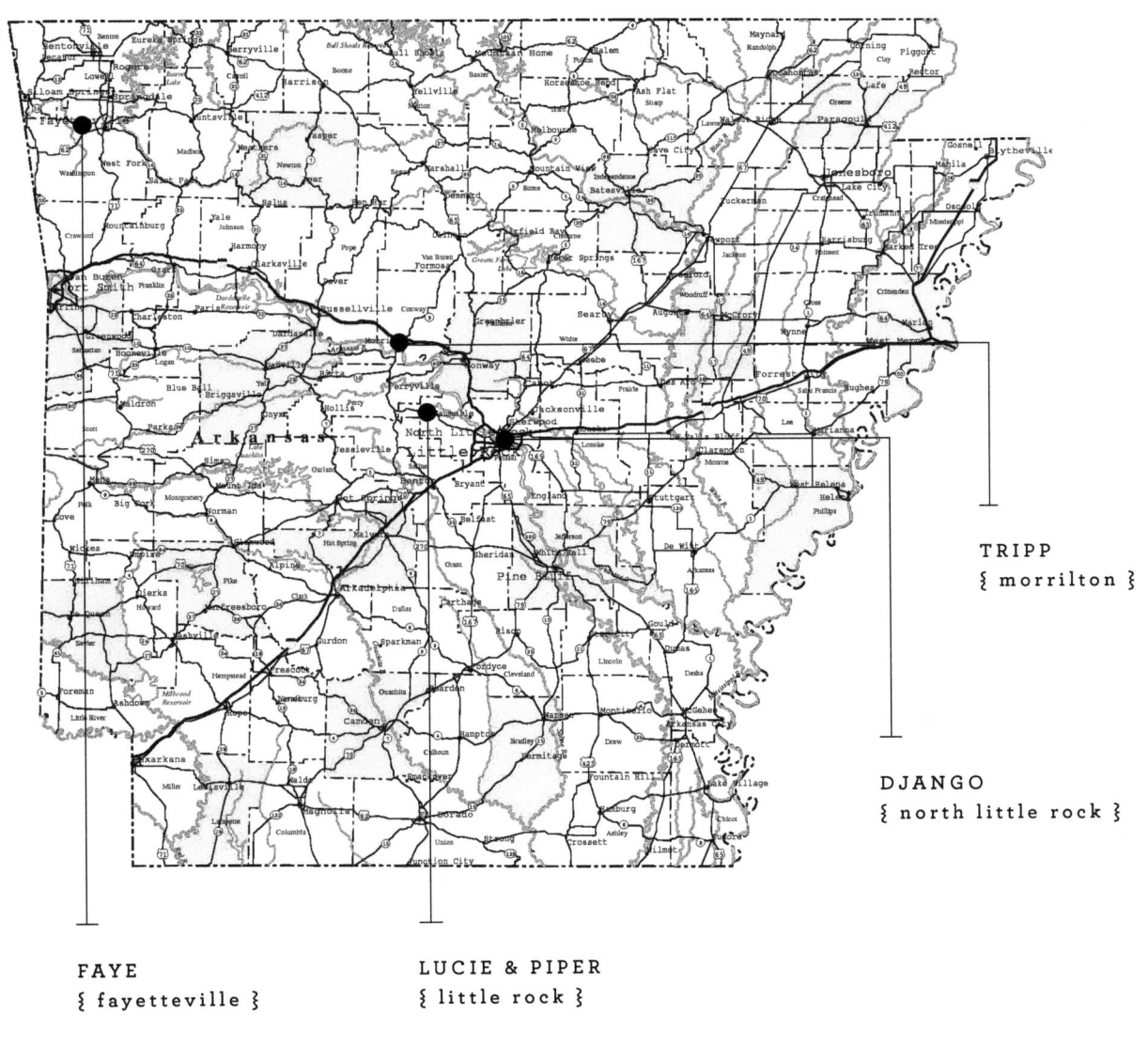

WHERE OUR DOGS WERE FOUND

TRIPP
{ morrilton }

DJANGO
{ north little rock }

FAYE
{ fayetteville }

LUCIE & PIPER
{ little rock }

ROCKY
{ new orleans, la }

{ ROCKY }

---

owners / judy & steve ostrowski

For all the hopelessness and devastation brought about by Hurricane Katrina, there were also unexpected gifts of help, love, and hope. Rocky is one of these gifts in a furry package.

Judy Ostrowski's husband, Steve, was deployed to Louisiana with the Arkansas National Guard for duty in the devastated parish of St. Bernard in Hurricane Katrina's aftermath. Judy was proud that as a family they could do their part to assist those affected, but she was worried about Steve's safety and was not looking forward to the weeks of separation from him. At the time Judy and Steve had two Yorkshire terriers; the two were 14 and 16 years old, and in their final years. Steve and Judy had discussed the fact that they should adopt another dog so that if one of their fur babies passed, the other would not be left alone. Their hearts were open to a smallish dog in need of a good home.

One night while Steve was in Louisiana, Judy received a phone call from him. He said he had a surprise to keep her company while he was gone, and that she would have to pick it up from the Little Rock airport. Then, he told Judy the story of meeting Rocky. While on National Guard duty, Steve was manning one of the checkpoints when an animal rescue volunteer, Mrs. Patty, came through the checkpoint with Rocky. Steve asked to hold the curly boy because he could see how frightened and devastated he was. Rocky was only four months old, a poodle/dachshund mix, and a living teddy bear.

Rocky's original family lost everything. They had to be rescued from their home and in the process voluntarily surrendered their puppies as they were evacuated because they could no longer care for them. They had put their two puppies in the attic of their home in hopes that the puppies could be saved at a later time. When the family was rescued, they told first responders about the two puppies in the attic, and the responders contacted animal rescue.

When animal rescue arrived, Rocky was found. Unfortunately, his sister and litter mate, Daisy, was not. In a plastic bag close to Rocky, the previous owners had left his "puppy book," along with his shot and vet records. From that book, rescuers learned that his name was Rocky and that Daisy was his sister. The family had written sweet things in the puppy book about Rocky and Daisy. It was clear that the family cared about their puppies and that having to leave them must have been devastating. Steve immediately told the volunteer, "I have a home for this dog!" He told Mrs. Patty, "My wife is the biggest dog lover ever, and has a soft spot for abandoned, homeless, stray, and disabled dogs. She has never met a dog she didn't love! This dog will have a home in paradise." He told her about their two aging dogs and the desire to help a dog in need.

Steve couldn't keep him in the tent at the checkpoint with the surrounding disastrous conditions, and Mrs. Patty lived in New York and was flying home the next day. Mrs. Patty promised Steve that she would "make it happen" and that somehow she would get Rocky to Judy. True to her word, Rocky flew to New York with Mrs. Patty. Rocky was then flown to another rescue volunteer in South Carolina who kept

him until he boarded his flight to Little Rock a couple of days later. Judy was worried that Rocky would be scared and sad if she was late to pick him up; she nearly was, having been involved in a fender bender on the way to the airport. On arrival, she told the airport cargo manager about Rocky and he said he would see what he could do. Everyone in line to pick up packages was kind and accommodating, making sure that Rocky and Judy were taken care of first. When Rocky was brought out in his pet carrier, he was obviously terrified. Judy bent down to his cage and said, "I'm your new momma. Do you want to go home with me?" Rocky, still frightened and shaking, put his paw through the cage and let her hold it. He has had Judy's heart ever since.

The Ostrowskis' two Yorkies passed away less than 6 weeks apart a few months later. Rocky did not replace them, but allowed Judy to focus on helping him to adapt to a new life and heal from the terrifying situation he had been through. Judy felt she was able to honor the memories of her Yorkies by nurturing Rocky.

His name has been changed to Rockford, but still known as Rocky, he now lives the grand life with Judy, Steve, and his 11-year-old adopted sister Lea Lea. Rocky is the best gift they have ever received, a shining soul. Rocky can perform a number of tricks and "speaks" when asked a question. Together, they participate in fundraising, pet races, and pet costume contests with various organizations to help homeless and abandoned dogs. Rocky has an enviable wardrobe and is a great sport about dressing up. He knows he looks adorable, and he owns it!

{ LUCIE } | owner / whitney bower

Looking at Lucie, you would never guess this beautiful queen snores louder than your average human.

Before her days of snoring in peace and comfort, Lucie was dumped on a very busy street in Little Rock and left to fend for herself. *Home Sweet Home*'s photographer, Whitney Bower, wanted to help when her friend found Lucie. Despite being discarded like garbage, she was still the sweetest, happiest little thing when Whitney met her later that day.

When Whitney took her to Hollywood Feed for a bath, she couldn't believe all the dirt and filth that washed into the tub. Her ears were matted and it was obvious that she'd recently had a litter of puppies. At the vet, where she received shots and tested positive for worms, it was determined Lucie was likely used for excessive breeding—mass producing puppy litters for a breeder's profit. Once the breeders couldn't use her anymore, they likely dumped her because she was worthless to them.

When Lucie found a home with Whitney and her family, they could tell she was never given simple dog "luxuries" like bones or toys. When Whitney gave her a bone for the first time, Lucie stared at it, not knowing what to do next.

It was love at first sight for both Whitney and Lucie, who is now the queen of the Bower household. For more than a year, she's had time to adjust to becoming completely spoiled.

{ TRIPP }

———

owner / anna serpente

There are no regulations on dog breeders in the state of Arkansas, which has lead to the proliferation of puppy mills in the state. Without regulations, no inspections are conducted, and many animals suffer. Arkansas is one of the top ten states nationwide in greatest number of puppy mills; it is estimated that there are anywhere from 700 to 1,000 in the state.

Tripp was born at a puppy mill in Morrilton in 2009. After the mill owners got into legal trouble for writing bad checks, they abandoned their mill and property, but not before locking the remaining dogs in an old school bus and leaving them to die. Luckily, the authorities had already been contacted about dogs on the property. Although some of the dogs weren't as fortunate, Tripp and several others were rescued from the bus.

Tripp landed at The Humane Society of Pulaski County. A friend of Anna Serpente, the woman who would become Tripp's mommy, knew that Anna had been on the lookout for a scruffy dog to rescue. When Anna saw him for the first time, her heart grew like the Grinch's. Tripp is Anna's "little boy," even though he is 115 pounds. Parents aren't supposed to have favorites, but Tripp is hers.

If you suspect a puppy mill in the state of Arkansas

call | 1-800-MILL-TIP

To help end puppy mills, follow

"ARKANSAS AGAINST PUPPY MILLS" | on Facebook

{ DJANGO } | owner / belinda rogers

Django was discovered in May of 2015 by Casey Carter, Director of Rock City Rescue in North Little Rock, as she left work one afternoon. Partially hidden between two garbage cans as he scrounged for food, she almost missed him. But it was Django's lucky day. As Casey approached him, she immediately smelled, then saw, the first indication of the life he'd been living. A chain, deeply embedded in his flesh, hung from his neck and reeked of infection. She managed to load him into her car and took him straight to a veterinary hospital where he had surgery to remove the metal tether. He also tested positive for heartworms, had demodectic mange, and was malnourished. His body would need lots of care and time to heal, as would the fear and distrust in his intelligent eyes.

In August of 2015, Django came to Belinda Rogers as a foster dog; this was his fourth foster home in three months. He'd had trouble getting along with other dogs—mostly because he could be a little overzealous in his attempts to play—and was haunted by his trust issues with humans, especially men. He had no social skills, was easily spooked, and had strikingly little desire for human interaction or affection. However, the Rogers had recently taken in a stray, Meathead, who was sweet and amicable, and he and Django bonded. Through Meathead, Django learned quickly—to sit, to take treats gently from hands, to sleep indoors, to play . . . even to once more wear a collar. And slowly, very slowly, he learned to trust Belinda. Several months into fostering him, Django lay on the couch and put his head in Belinda's lap for the first time. It was monumental.

For all the strides they were making at home, Django's people anxiety made public adoption events torturous. At one, he pulled away and hid under Belinda's car. And his health issues continued. A "fast-kill" heartworm treatment wasn't conclusive, and it also caused a flare-up of the demodectic mange. Six months into fostering, Belinda approached Casey with concerns about Django's potential to be rehomed, and Casey agreed that what was best for Django was to stay with Belinda.

Belinda formally adopted Django in February of 2016, and has no regrets. He is still on a slow-paced treatment for heartworms that they hope will completely clear soon, but he's otherwise strong and healthy. His bond with Meathead endures, as has his comraderie with the Rogers' two other rescue dogs. They've even fostered and placed two additional dogs since adopting Django, and he has been a wonderful role model and playmate to them. He is still learning to trust strangers, but has come a long way in just about a year and half. He may never be what he could have been without such horrific abuse, but he's already so much more than he was when Casey found him scrounging between garbage cans.

Today, Django is finally able to focus on receiving love and giving hugs.

{ PIPER }

owner / anna coon

Piper was picked up as a stray. She was said to have a home, but she wasn't well taken care of. As may have been the case with Piper, not everyone is made to handle the responsibility of pet ownership, and that is okay! The tragedy is when people adopt an animal not realizing the requirements and then don't tend to the animal's needs, such as health issues. Before we consider adopting, we should all ask ourselves whether we can be responsible for another's life, providing them with everything they need—financially, emotionally, and otherwise. Dogs become family members and owners must be 110 percent invested.

Driving down a street in Little Rock during a rain storm, Anna Coon said a prayer: "Please God, show me what you want me to do with my life." Seconds later, a little dog ran across the street. Anna stopped and approached the dog, who literally jumped into Anna's arms. Anna tied to find the dog's owner and couldn't, so she took her up the street to a vet. She was skin and bones, and had no microchip.

Anna called her then-boyfriend, now-husband, and said she had a surprise for him. They already had three dogs and he knew Anna wanted a fish. When little Piper ran up to him, he told Anna, "You did the right thing." The rest is history. Although she was spayed, they quickly learned just how poorly Piper had been taken care of. Since the Coons took her in, Piper has continued fighting a skin infection. The vet couldn't tell how old she was, but guesses around eight years. A dog DNA test revealed she is mostly Chihuahua mixed with some other breeds.

Piper doesn't care for other dogs, but loves to snuggle with humans. If you don't pet her, she will put her head under your hand until you do!

{ FAYE } | owner / alex graham

Did you know black dogs, and cats too, are euthanized at a higher rate than other animals? They also remain in shelters much longer. The phenomenon is so common that it was given a name: Black Dog Syndrome.

Around 10 years ago, Alex Graham accompanied her college roommate, Lindsey, to Petsmart in Fayetteville, where a local shelter brought puppies every other Saturday. Lindsey was looking to adopt a puppy, and Alex tagged along for moral support. When they found the puppy pen, Alex immediately pointed to a little black dog sitting quietly in the back while the rest of the puppies jumped up, barked, and clambered for their attention. The shelter had named the quiet black dog "Fudgie." (Alex told them that was a terrible name, especially because she was not brown.) The staff told the girls that when "Fudgie" was just a couple of weeks old, someone had used a shoestring leash to tie her to a pole outside the shelter in the middle of the night. Left out in a bad thunderstorm, she wasn't found until the next morning. She was underweight and terrified.

After Lindsey spent a couple of hours cuddling the puppy and being incredibly indecisive, Alex decided there was absolutely no way they were leaving without her, and decided to keep the pup for herself. "Fudgie's" adoption fee was the best $35 she has ever spent.

Now "Fudgie" (renamed Faye) spends her days being a lazy old lady and foster sister to 32 pups and counting, their ages ranging from 5 weeks to 3 months. By any standard, she is officially the most patient dog ever.

{ PART III }

"Patience, love, and frustration were daily feelings for us. Slowly, he felt more and more comfortable. We wish that he had always been with us. We wish we knew his whole story, but that is not possible. Instead, we love him and enjoy his quirky personality, feeling so blessed that he is with us now."

/ susan foster, RILEY

WHERE OUR DOGS

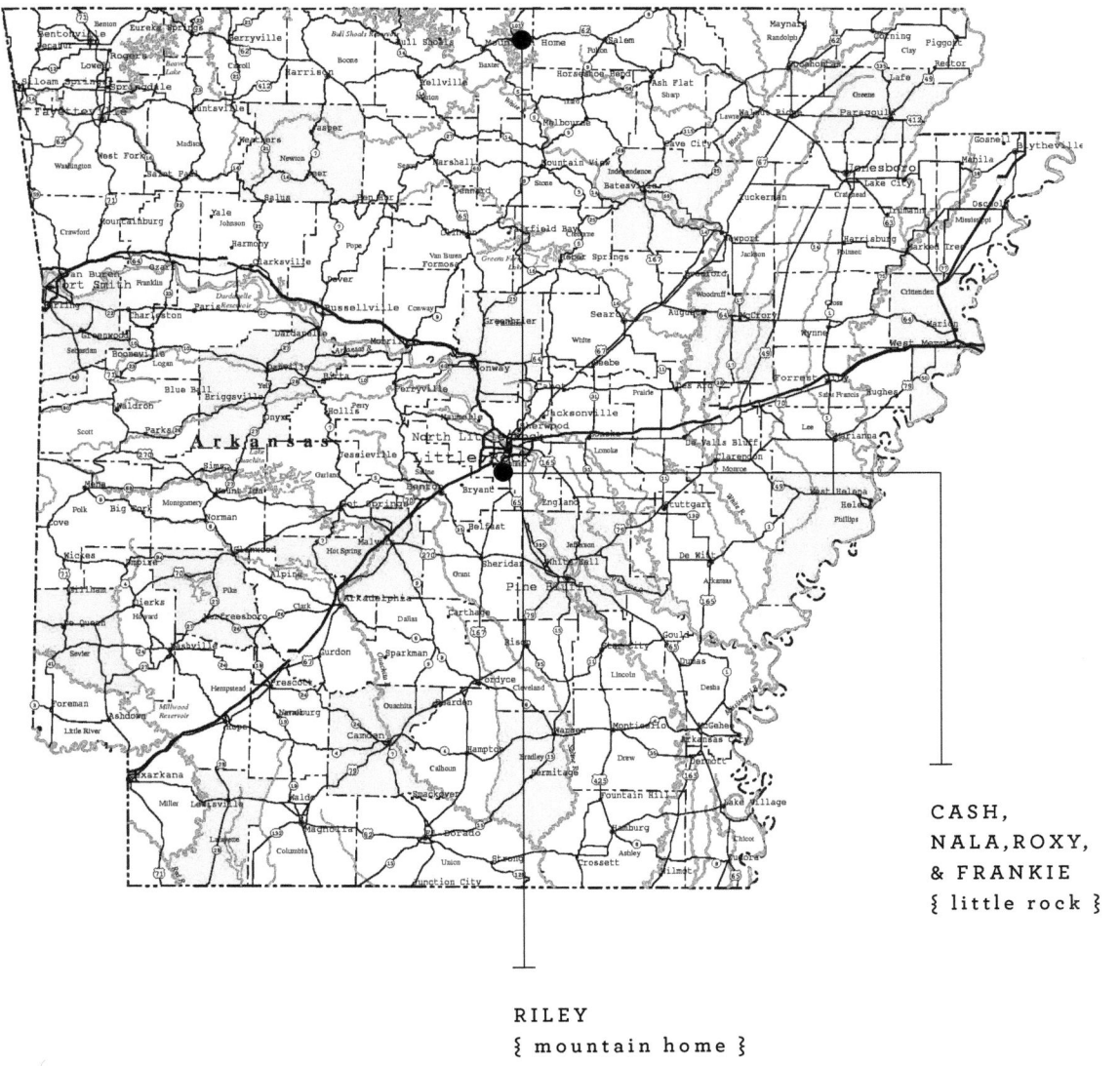

CASH, NALA, ROXY, & FRANKIE
{ little rock }

RILEY
{ mountain home }

WERE FOUND

{ CASH } | owners / donnie & meaghan ferneau

Cash Ferneau was adopted in 2008 from Central Arkansas Rescue Effort (CARE). Before the shelter, Cash is believed to have been homeless and abandoned. He was found in a dog park, beaten within an inch of his life. His legs were broken, his ribs broken and cracked. He had a collapsed lung, a cracked skull, and wounds on his side from a BB gun. Once rescued, the vet found a terrible ear infection, making it so he could barely hear. He was left in a no kill shelter for over a year with no one interested in adopting him. (He is also a black dog, which didn't help his cause.)

Donnie Ferneau walked into the shelter one day and saw Cash. Named Tune at the time, he was so scared that he cowered in a corner and peed on himself. The shelter told Donnie that he did this every time someone walked by, which is probably why he hadn't been adopted for over a year. So naturally, Donnie took him home.

Cash didn't understand the concept of toys, and really didn't know how to "be a dog." For over three months, it was nearly impossible to find Cash in the house. He was so terrified that he wouldn't eat when food was laid out for him. Eventually, Donnie and his wife Meaghan discovered Cash had created his own burrow under the bed. He saved things like candy, socks, and other little trinkets from around the house. One evening, Donnie decided to sit on the floor with his and Cash's dinner in hand, eating until Cash came out of hiding. After that, Donnie couldn't shake Cash. The two have been attached at the hip ever since.

Shortly after the adoption, Donnie went through a terribly rough part of his life. Cash was the only constant, and they created a bond that Meaghan has never witnessed elsewhere between man and dog. The Ferneaus believe that Cash saved Donnie.

After some years, Cash became more "dog-like." He started to understand that toys were for play, that the dog park is actually a fun place, and that he has a family that loves him to the moon and back. He still saves his favorite things under his bed—mainly the squeakers from inside toys, which he carries around without tearing. The Ferneaus like to call the squeakers "Cash Bucks" since they are his prized possessions. Though a little older and a little whiter, Cash is the best dog anyone could ask for. He's obedient, but more importantly, he's sweet, he knows when family members are not well, and he loves so hard. He is known for giving hugs by wrapping his paws around his humans' shoulders while standing on his back legs.

{ NALA & ROXY }

—

owner / olivia jones

Nala was found running in the woods in Little Rock with a fox trap closed on her paw. The family who found her immediately took her to the vet. For several days, she underwent treatment for her injured paw and for malnourishment. She was approximately ten months old. After the vet verified she did not have a chip, her picture was posted on several lost and found websites. No one claimed her. Olivia Jones took her home and named her Nala.

Olivia attempted to kennel train Nala. She came home from work at lunch time the first day and discovered that Nala had managed to break out of the metal cage. (Olivia still cannot figure out how Nala did this as the bottom does not slip out and the kennel door was locked.) Nala had seemingly disappeared. Olivia finally found her standing on the roof; she had unlocked a window and jumped out. After this incident, Nala was diagnosed with severe separation anxiety and had to go to doggy daycare.

Nala also had a large spot on her back that appeared to be mange, although she never tested positive for it. After various treatments, Olivia realized this spot was

spreading to Nala's face. The vet suspected that she did have mange but was not testing positive because the mites were so deep in her skin. With treatment, her mange finally cleared up.

Nala now lives with an older rescue boxer, Roxy, also adopted by Olivia. After being bred, Roxy was found left tied to a tree with no food or water in the searing heat of the Arkansas summer. She was diagnosed with pinworms, hookworms, and heartworms so severe that Olivia and her husband were not sure she would survive the treatment. She was severely emaciated and dehydrated. Approximately 10 years old, Roxy is now treated like a queen, and could not be a more faithful, loyal, and loving companion.

Nala and Roxy are the best of friends, and find comfort in companionship with each other. One of their favorite things to do as a family is floating the Buffalo River. They love riding on their canoe!

{ FRANKIE } | owners / arny & sabra ferrando

If someone told you that a dog found in a shopping center parking lot in Little Rock could detect cancer, would you believe them? Dr. Arny Ferrando, PhD, and his wife, Sabra, were shopping when they noticed a shepherd mix in poor health and searching for food. Sabra is a volunteer at the Little Rock Animal Village, so she always has dog treats in her car. She called for the stray, and he jumped in their open car door in hopes of being fed.

Soon enough, he was named Frankie and became a member of the Ferrando family. He quickly shared his love of being a working dog and going on adventures. These drives made Frankie the ideal fit to become a certified search and rescue dog. Being a search and rescue dog required knowing how to signal for a certain scent when he detected it. Because of his own professional background and training, Dr. Ferrando knew Frankie was special and began training Frankie to do something that no other dog has done before: smell out cancer.

Frankie is the first dog trained to detect thyroid cancer with a sniff of his nose. Thyroid cancer is relatively rare. At present, diagnosis of thyroid cancer is tricky, and standard clinical tests are often inconclusive. Throughout six months at University of Arkansas for Medical Sciences (UAMS), Frankie was trained to differentiate between urine, blood, and thyroid tissue obtained from patients with and without thyroid carcinoma. Frankie was taught to alert (lie down) if the sample was positive, and turn away if it was negative. After this training, he was asked to help "diagnose" the presence of thyroid cancer in clinical patients. He successfully diagnosed 30 out of 34 samples, a remarkable 88 percent success rate.

Frankie is now retired from his detection career and living the spoiled life, as any dog should. He is the founder and namesake of the Frankie Foundation, which is devoted to funding groundbreaking research that involves using dogs to detect human cancers.

{ RILEY }

owner / susan foster

In early December 2012, a rescuer was contacted by a young woman about her younger sister's dog. The girls' mother had threatened to shoot the dog, and they didn't know what to do amidst the crisis.

The younger sister had rescued the dog from a shelter where she volunteered. He was picked up as a stray and was soon to be euthanized. Familiar with shelters, he'd had three shelter stays with adoptions but had not managed to find his forever home. The older sister pulled together the adoption fee and took him home. The younger sister did the best she could. The girls' mother had several pit bulls with whom her dog, then named "Bosco," had to compete for daily meals. To feed the dogs, the mother tossed food scraps out of the back door. It was a fight to get anything. Bosco suffered many bite wounds from these scuffles. He roamed the surrounding farm lands. He was skiddish around strangers and didn't trust many people, especially men. The mother didn't like Bosco and wanted him gone. She threatened to shoot him. The older sister contacted a rescue group through Facebook. The group posted an appeal for a volunteer in the area to help Bosco before her mother acted on the threat. Not long after, the sister learned her mother had paid a man to shoot Bosco. The would-be assassin did shoot Bosco, but Bosco wasn't dead; he'd managed to run off. Now, the search was on for a wounded dog.

A woman in central Arkansas volunteered to search for him alongside the older sister. For four days they called out and watched with hope. On the fourth day he was found. Bosco had been hit in the right hind leg. Unfortunately, this wasn't his first encounter with a gun. He still carries shrapnel in his left flank and his chest as proof of this violent history. When the gun was fired this time, Bosco must have remembered the sound and understood that it was time to bolt. His femur was shattered, his leg infected. The volunteer rushed him to her vet near Little Rock where his wound was cleaned and his infection treated. When he was stable and able to travel, he was transported to Rocky Ridge Refuge near Mountain Home. Owner Janice Wolf cared for him until he could have surgery on his injured leg. It was doubtful that the leg could be saved, but the vet decided to go forward with the operation. The shattered bones were pieced together with a large plate.

This is when Susan Foster was contacted about Bosco. Janice from Rocky Ridge asked Susan if she could foster Bosco, now renamed Jingle, while he recovered from his surgery. On January 4, 2013, Jingle made his way to Susan's home to complete his recovery. Jingle was still covered in bite marks in various stages of healing. Susan brought him into the backyard to meet her other two dogs, Trixie and

Molly. In the days to come, Jingle made his home in Susan and her husband's hearts. They knew he would never leave. Jingle became Riley, and found his forever home.

Riley suffered from severe anxiety—almost everything would make him nervous. His leg slowly healed and he was treated for heartworms. Months passed. Riley was no longer limping, and was bearing weight on his injured leg. They watched him begin to shed his fear, and trust again. Riley is now home for good, and no longer has to worry if he is going to wind up at a shelter or looking down the barrel of a gun.

{ PART IV }

"Sometimes I look at her and I just cry because I know she was meant to be part of my family."

/ lindsay newsom, LYLA

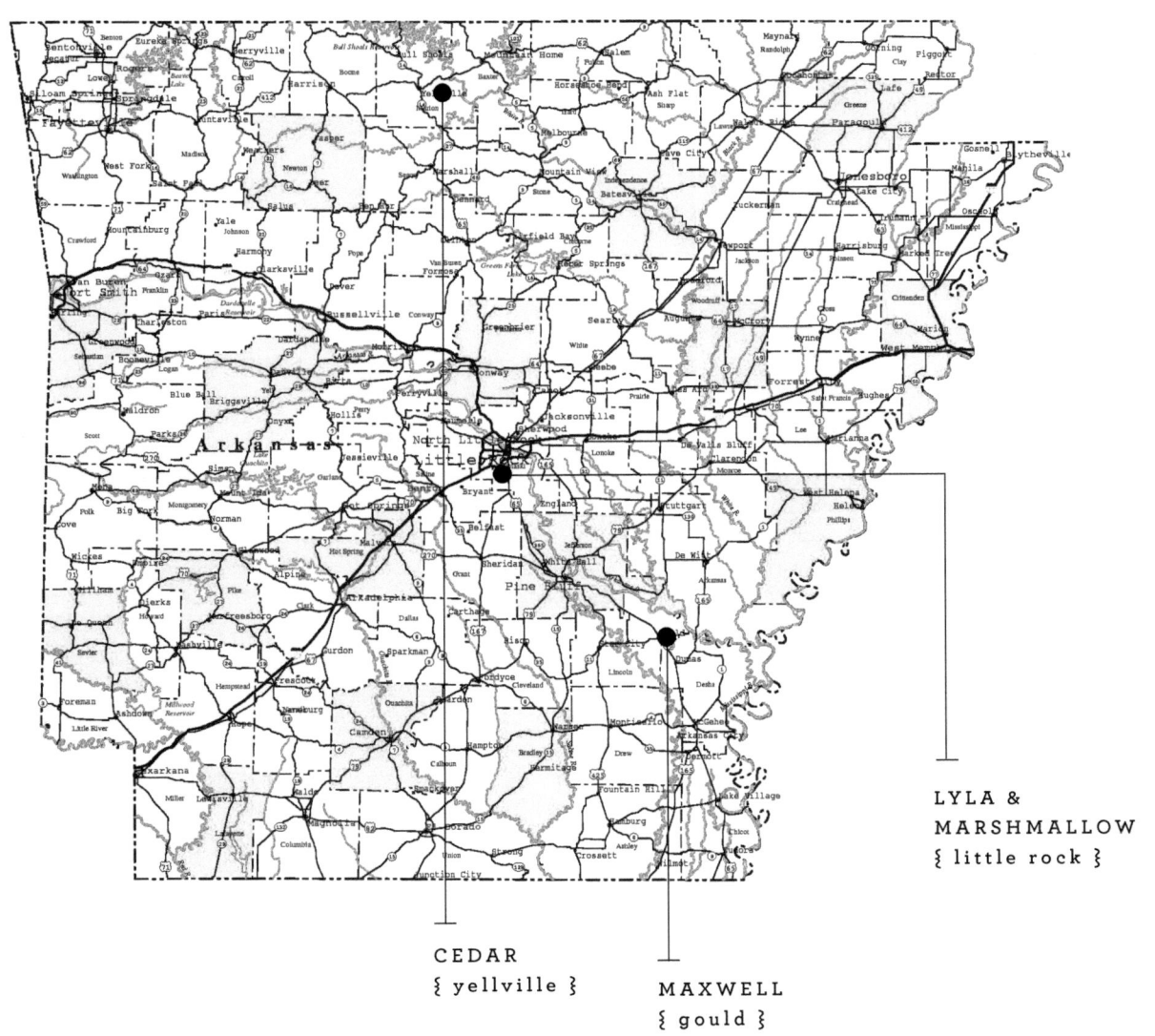

{ MAXWELL } | owner / priscilla gray

On January 29, 2016, snow blanketed south Arkansas. A lab-hound mix puppy was found on Highway 65 around Gould with two broken hind legs. He had been beside the road for two days, while cars continued to drive by. Finally, a lady from Lake Village stopped. The lady from Lake Village took the puppy to a large-breed vet, who splinted both back legs—a temporary fix at best. She reached out to a group of search and rescue folks in Little Rock. Through this group, Priscilla Gray saw a photo of Maxwell and managed to get involved.

Maxwell needed surgery immediately. Priscilla called her vet, Dr. Laura Mehaffey at Pinnacle Valley Animal Hospital, to see if she would be willing to help Maxwell. Dr. Mehaffey volunteered her clinic's services, and Maxwell was brought to Little Rock. Incredibly, Dr. Mehaffey paid a veterinary orthopedic surgeon to do the surgery on Maxwell. Priscilla met Maxwell for the first time at Pinnacle Valley. He had her at first sniff. Priscilla visited him every other day and on Sundays for six weeks. After six weeks, Maxwell finally got to "go home" to Priscilla's house for the first time and begin rehab. Fortunately, Priscilla's other dogs accepted Maxwell into their home.

Maxwell has come a long way, but still cannot sit down (even though he doesn't realize that!). Following his surgical recovery, the next challenge was treating his heartworms. This was Priscilla's first experience with canine heartworms. Heartworm treatment is a strict and expensive program, so it is key to follow protocol precisely once a diagnosis is made. She wanted to stop mid-way through the treatment, afraid that it was hurting Maxwell, but she never gave up.

Maxwell has recovered from heartworms and continues to improve, but the road has been rough for him. Priscilla knows that to rescue a dog like Maxwell takes a team effort, and she could not have done it without the generosity and kindness from Dr. Mehaffey and the staff at Pinnacle Valley Animal Hospital.

{ LYLA }

———

owner / lindsay newsom

Lindsay Newsom's favorite quote is, "My shelter dog rescued me."

When she thinks back to her earliest memories, they include shelter dogs Willie and Andy. Lindsay's family adopted Willie from The Humane Society of Pulaski County in 1979. He was Lindsay's first true friend. Willie taught her to share, love, and let go when the time was appropriate. Andy was her family's second rescue from The Humane Society, adopted when she was 16. He became her car riding buddy.

Andy was a sweet and happy soul. Her family lost him when Lindsay was pregnant with her first child.

When Lindsay was in college, she adopted her very own dog, Duke, from The Humane Society of Pulaski County. He was her protector; as long as Duke was around, Lindsay was safe. He loved her like no human could. Three years later, she adopted Maple. Duke and Maple were a huge part of her life. If you knew Lindsay, you knew Duke and Maple. They saw her through happy times and tough times: through college, graduate

school, her first job, a divorce, finding love again, and babies. When Lindsay lost Duke in 2011, Maple saw her through. Maple went to be with Duke in 2015. As another part of Lindsay's heart was forever gone, she found comfort knowing that Duke and Maple were together again.

For over a year, Lindsay looked for another rescue dog, but none fit her. Not until Lyla came along, rescued from Out of the Woods Animal Rescue in Little Rock in January, 2016. Lyla is as intelligent as she is gorgeous, and Lindsay is training her to be a therapy dog. This training has also been great therapy for Lindsay, helping her bond with Lyla. Lyla has shown Lindsay that after the loves whose loss you will never get over, there's still life.

{ MARSHMALLOW }

---

owners / emese & collin boone

Many towns have dogs that everyone knows by name or that become emblematic of the community. For more than a decade, Marshmallow has been the "shop dog" at Box Turtle, a gift and clothing store located in the Hillcrest neighborhood of Little Rock. Box Turtle's owners, Emese and Collin Boone, have always been avid rescuers. In fact, they are known for matchmaking families around Little Rock with their perfect rescue.

Anyone who has visited Box Turtle has most likely seen Marshmallow walking around or enjoying a snooze somewhere in the store. Possibly, this little guy has even encouraged you to pass him a treat or give him some lovin'. Marshmallow now rotates his Box Turtle duties with his two brothers, Mash and Moose.

The unique thing about Marshmallow's story is that there isn't really a story; his past remains a mystery to this day. He was found outside of Leo's Restaurant in Hillcrest with no collar or microchip. He was trying to jump into a Jeep outside the restaurant when Emese, Collin, and their son walked by. The family in the Jeep didn't know where or to whom he belonged, so Emese and Collin decided to keep him at their house until the owner surfaced. Their son was very young at the time and mentioned, "He sure does

look like a marshmallow." Like its inspiration, the name stuck.

With no leads, Emese and Collin decided to let Marshmallow hang out on their front porch in Hillcrest so that if the owner walked by they would hopefully see him. No luck. Soon, they decided they would let him hang out at the store with hopes of the owner walking in. No luck. No calls from a vet or a shelter with news about his owner.

The Boone family had another well-known shop dog named Cheeky, who was also a rescue. He liked being a loner dog and wasn't fond of having "dog company." Marshmallow figured out how to stay out of Cheeky's way long enough to finally be accepted.

As time went by, the Boone family became attached to Marshmallow, and ten years later, they're so glad that his owner never wandered by their porch or into the store to reclaim him. Marshmallow's past will likely remain mysterious, but he has found his home sweet home with the Boones and among so many in the Little Rock community.

{ CEDAR } | owners / angie & bill sheehan

On a winter day in Yellville, one-year-old Hazel showed up on Angie Sheehan's porch. Exhausted, starved, and very pregnant, remembering the state Hazel was in still brings tears to Angie's eyes. Hazel seemed like she was ready to give up on life. Because Yellville is so small, Angie learned more about Hazel's past; she had been taken in as a puppy by a teenager who never properly cared for her. Hazel either ran away or was dumped like trash because she was pregnant.

Angie and her husband Bill decided to let Hazel stay, rest, and enjoy her puppies when they came. Hazel gave birth to twelve tiny puppies. Only six of them lived. Hazel was horribly malnourished, and at first produced no milk. There was a point when Angie feared all the puppies would be lost.

Despite the life she'd led, Hazel was a sweet, loving dog, and flourished with attention and lots of food. Still, Hazel was scared of Angie and Bill's then 4-year-old grandson, which worried them. They understood this was the result when dogs aren't socialized as puppies. Angie knew it would be a challenge to find Hazel a new home since she was scared of children, but they were determined to find good homes for her, the three female pups, and the three male pups. Meanwhile, Hazel took excellent care of the puppies and they grew quickly. Angie's mind was set that the puppies were temporary and they would all go to loving homes, but she relished seeing their progress.

Cedar, a puppy from Hazel's bunch, was special from the beginning: the first to figure out how to jump out of the dog box, the one interested in exploring the living room when the others cowered in a dog pile, the one who would throw the biggest fit at supper when they were old enough for dog food. Cedar was what one might call "the pick of the litter."

It was a small miracle when an awesome dog rescue organization, Gail's Pets Second Chance, said they'd take Hazel and her babies. Then Angie realized she couldn't let them all go—she had to keep one to love. Their hearts had been captured by little Cedar.

Cedar is now the center of the Sheehan's world. She has grown into a funny, beautiful, faithful dog. Her brothers and sisters found good homes. Their mommy, Hazel, was renamed Smoky and sent to Massachusetts where she lives the happy dog life she deserves with a Harvard Law professor. Thanks to the Sheehans and Gail's Pets Second Chance, six dogs now have loving homes.

{ PART V }

"He clings to me for safety, and I cling to him for comfort. Adopting him was one of the best decisions I could have made. He really is my best friend."

/ reagan boykin, TEDDY

WHERE OUR DOGS

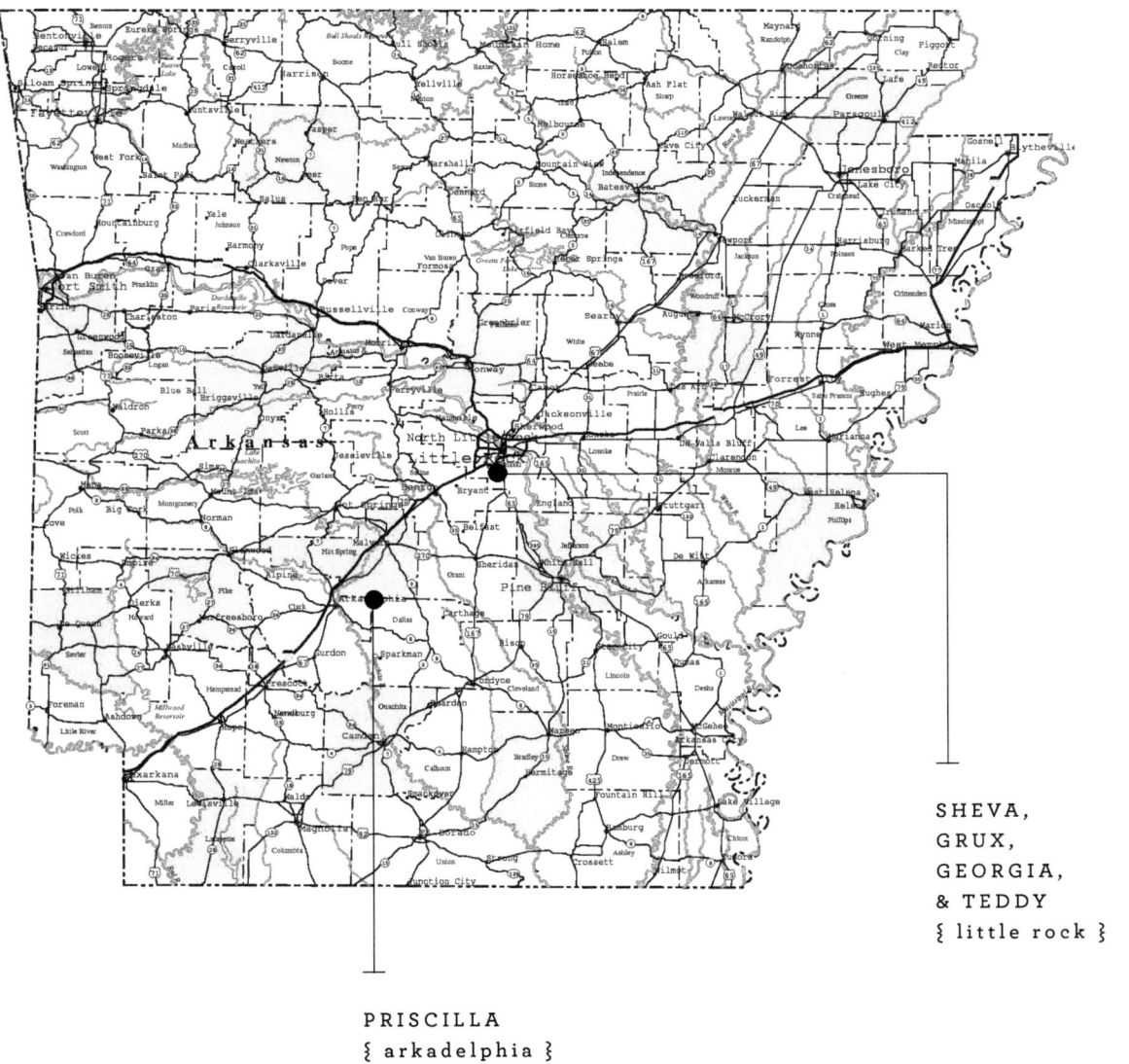

SHEVA,
GRUX,
GEORGIA,
& TEDDY
{ little rock }

PRISCILLA
{ arkadelphia }

WERE FOUND

{ PRISCILLA } | owner / rennie karnovich

Meet Priscilla Louise. She was found in a box beside the highway in Arkadelphia when she was about 10 weeks old, and taken to the Clarke County Animal Shelter where she was named Sugar. On January 4, 2003, when "Sugar" was 14 weeks old, she was adopted by Rennie Karnovich, and renamed Priscilla Louise.

Priscilla was Rennie's first dog as an adult with a full-time job, and she quickly became spoiled with collars and leashes, toys, doggy play dates, and dog birthday parties to rival any toddler's. She went to puppy training, obedience school, and more. Rennie did everything she could with Priscilla.

One day, Rennie and Priscilla went to a snow cone stand to enjoy their favorite root-beer-float-flavored snow cone. A lady at the stand asked Rennie if her young child could pet Priscilla. "Of course! Priscilla loves kids." Rennie asked Priscilla to sit. The little girl approached quickly and stuck her thumbs up Priscilla's nose. This definitely could have been a very bad situation for all involved, but fortunately Priscilla pulled her head away and looked at Rennie as if to ask, "What should I do now?" Rennie understood what an incredible response this was from a dog—looking to her owner for cues as to what to do next in an unclear situation.

About a year later, Rennie began a new job at Arkansas Children's Hospital. She saw a dog going into the building and asked why the dog was allowed in the hospital. She was told the hospital had therapy dogs that greeted the children. It was the first time Rennie had heard of such a thing. Priscilla's behavior with the girl at the snow cone stand had already indicated Priscilla was a natural with kids. Rennie researched and found a great dog trainer to help them sharpen skills; Priscilla worked on responding to wheelchairs and weird sounds, and Rennie learned how to always be an advocate for Priscilla.

Rennie and Priscilla passed their test, and became a therapy dog team working at Arkansas Children's Hospital as part of the Therapeutic Animal Interventions Lift Spirits (TAILS) program. After 12 years of volunteering, they have tons of stories they will never forget. One favorite memory was of a tearful family looking on as, following surgery, their daughter took her first steps to hug Priscilla. A framed photo of Priscilla and the girl remained next to her bed throughout her hospital stay.

Priscilla's trademark—other than doing cute tricks like hide-and-seek, giving high fives, saying her prayers, and whispering—are her painted toenails. For every visit, she has a pretty "pawdicure."

In June 2016, Rennie moved Priscilla and her two other rescue dogs to Phoenix, Arizona, so Priscilla could take on a new frontier, Phoenix Children's Hospital.

{ SHEVA } | owner / denise palmer

Denise Palmer and her daughter went to Central Arkansas Rescue Effort (CARE) in Little Rock with hopes of adopting a cat. On arrival, they discovered that the cat they had become interested in online had already been adopted. As they were about to leave, a gentleman came walking up the sidewalk with a beautiful black and white, medium-sized dog. The dog seemed so scared when the man carried her into CARE.

The man, Shane, told the story of how he and his 16-year-old daughter were practicing driving in a south Little Rock parking lot when they saw this dog by a dumpster. Shane had gotten out of the car and sat on the curb near the dog, which eventually came up to him. Shane pulled ticks from her coat, and she seemed thankful. With the urging of his teenage daughter, they brought the dog home with them. They kept her for a few weeks until his daughter had to go back to Hawaii and he had to start traveling for work. Shane had no family in Little Rock, and therefore no one to help with the dog. Denise and her daughter listened to his plea to find the dog a home.

The dog's sweet brown eyes met Denise's. She immediately agreed to foster. It appeared as if the dog had either had a litter of pups or was possibly pregnant, and everyone involved knew this would make an adoption very difficult. Denise took her to the vet the next day to discover that she was indeed pregnant. The vet thought she was gestating five pups.

Denise was urged—not by the vet but by another organization—to have the pups aborted. It was thought that the pups probably wouldn't be healthy due to their mother's poor health. Denise scheduled the abortion and showed up for the appointment. She was sick about it. She couldn't let them perform the abortion. Even if only one pup lived, Denise decided it was worth the risk. The vet techs stood with Denise against those that wanted her to abort the puppies. A few weeks later, she birthed 9 pups: All were healthy!

With the help of CARE, friends, and coworkers, homes were found for all 9 pups. Denise's family decided they loved the mama too much to let her go, so they adopted her and named her Sheva. They cannot imagine their lives without her. Sheva brings joy to their family each and every day.

{ TEDDY }

—

owner / reagan boykin

Stereotypes are common when discussing dogs, and certain breeds experience a lot of discrimination. Breed-specific legislation makes owning some breeds illegal. Many people automatically think of pit bulls and Rottweilers as "dangerous," and other breeds including chow mixes have reputations for their so-called "bad temperaments." Chows are often in shelters for long periods because of stereotypes, and many potential owners don't like the idea of having to spend the extra time necessary to train a chow mix.

Teddy is a chow mix who was surrendered by three different families. The longest he had ever been with a family was ten months. Teddy sat in the shelter for over a year waiting on someone to adopt him. Due to trauma and abuse that he endured at his first home when he was a tiny puppy, it made it extremely difficult for people to open up their homes to him because he was very cautious about letting people "love him." He had hip and leg surgery when he was a puppy, and as a result needed daily medications. Teddy couldn't walk or run up and down flights of stairs easily due to his surgeries. People didn't like the idea of having a "high maintenance dog" that needed a lot of extra attention.

His owner, Reagan Boykin, adopted this sweet boy from The Humane Society of Pulaski County in February of 2016. Reagan and Teddy's first couple of days together were a little rocky. Teddy had to learn to trust humans again, and Reagan had to show him that she wasn't going to hurt him or send him back to the shelter. He needed time to understand that she was going to love him for good. After all, the longest a family had committed to him was ten months.

Reagan occasionally has severe panic attacks. One day she was having an attack and Teddy jumped up in the bed with her, letting her hold him until she calmed down. If you know anything about the chow breed, you understand this rarely happens, especially with a shelter pup that has experienced only pain and suffering.

Teddy now goes to work with Reagan two or three days a week. They go practically everywhere else together, and especially enjoy attending softball games. In Reagan's mind, Teddy saved her; Teddy knows Nicole saved him.

{ GRUX }

owners / kevin & sarah o'connor

Meet Grux (pronounced "Gr-ucks"). His family named him after LeRoi Moore, saxophonist and founding member of The Dave Matthews Band, better known as "The Groo-Grux King." Grux has not always been his name. Two years ago, the animal shelter named him "Berry" because it sounded better to prospective families than "citizen surrendered stray #4765 from Little Rock, Arkansas."

In autumn 2015, Grux was found roaming a neighborhood with no collar, and seemingly no family looking or hoping for his return. A kind citizen scooped him up and dropped him off at the local animal shelter. He spent four and a half months living at the Little Rock Animal Village waiting for his old family to find him or a new family to choose him as their own. Days turned into weeks and life in the shelter became normal. Grux was later moved to The Humane Society of Pulaski County. The staff got him settled into a kennel and gave him one of his favorite things—a brand new tennis ball.

On February 28, 2015, Kevin and Sarah O'Connor arrived at the shelter with the intention of just looking around. Little did they know they would be charmed by Grux's tranquil demeanor and poise around other

dogs and cats. They saw him playing with his tennis ball and immediately feel in love. After they played and went for a walk together in the yard, they knew they'd found the outdoor adventure partner they'd been searching for. Adoption papers were signed and they took their first family photo.

Kevin and Sarah celebrate Grux's birthday every year on February 28, the day his name was changed from "Berry" to Grux, the day their adventures began as the Three Musketeers. When it is nice outside, the trio goes camping in the Ozark Mountains. Grux supervises them from his perch as Kevin and Sarah rock climb. Camping is second on Grux's list of favorite things, right behind his tennis balls.

Kevin and Sarah never stop wondering where Grux came from or how old he really is. Mostly, though, they are thankful for where he finally ended up, snuggled between them in the tent when they go camping or greeting them at the door after a long day at work. Grux even got the honor of carrying their wedding bands down the aisle on their wedding day.

{ GEORGIA } | owner / laura daubous

Laura Dabous' first fostering experience was through Rock City Rescue in North Little Rock. Her first foster dog, Delta, was found in southwest Little Rock, having just given birth to eight puppies. Starving and homeless, the poor girl had her babies under a house in the oppressive Arkansas heat. Positive for heartworms and tapeworms, covered in ticks and fleas, she also was covered with flesh wounds.

Rock City Rescue also rescued Delta's eight puppies, all now in loving homes. Her pups are doing great, and Laura keeps up with a number of them. One of them, named Nola, hiked the third highest peak in the US, Mt. Foraker in Alaska. Another puppy, River, sensed a woman in trouble and pulled his human mom in her direction to help. A third puppy named Jett has learned to fetch the morning paper for his family. A few of the pups went to homes where the families had lost beloved pets, so the puppies are helping to fill up their hearts again.

While Laura was fostering Delta, Rock City Rescue had her treated for heartworms. During those months, the Dabous family came to know that Delta was part of their family. They officially adopted her in December of 2015, and she became Georgia Mae. She is now healthy and heartworm free. She weighed just under 50 lbs when she was rescued, but now tips the scales at a whopping 73 lbs!

Laura thinks that Georgia is a true princess, who just never knew it until the Dabous family came along. Georgia quickly adapted to the spoiled life, and has a whole couch to herself, along with pretty much everything she desires. She won't go near the outdoors if bad weather is looming or if there is a drop of rain —she can't stand to get her paws wet—and the Dabous family is happy to accommodate her. Georgia Mae loves to be brushed and pampered.

RESOURCES

/

select animal rescues of arkansas

+

{ ALFA } - a love for animals
facebook / @aloveforanimalrescue
location / jonesboro

+

{ ARKANPAWS ANIMAL RESCUE }
website / arkanpaws.org
facebook / arkanpaws animal rescue
location / conway

+

{ ARKANSAS PAWS IN PRISON }
website / adc.arkansas.gov
facebook / @arkansaspawsinprison
location / little rock

+

{ A & B } - animal rescue of southwest arkansas
facebook / @abrescueofswarkansas
location / ashtown

+

{ CARE } - central arkansas rescue effort for animals
website / careforanimals.org
facebook / @careforanimals98
location / little rock

+

{ FAYETTEVILLE ANIMAL SERVICES }
website / fayetteville-ar.gov/525/animal-services
facebook / @fayettevilleanimalshelter
location / fairfield bay

{ FUZZY HEARTS ANIMAL RESCUE }
website / fuzzyheartsrescue.yolasite.com
facebook / @fuzzyheartsanimalrescue
location / little rock

{ HEBER SPRINGS HUMANE SOCIETY }
website / awos.petfinder.com/shelters/AR56.html
facebook / heber springs humane society
location / heber springs

{ HUMANE SOCIETY OF THE DELTA }
website / humanesocietyofthedelta.org
facebook / @humanesocietydelta
location / Helena

{ HUMANE SOCIETY OF THE DELTA }
website / humanesocietyofthedelta.org
facebook / @humanesocietydelta
location / Helena

{ HUMANE SOCIETY OF FAULKNER COUNTY }
website / fayetteville-ar.gov/525/animal-services
facebook / @fayettevilleanimalshelter
location / fairfield bay

{ HUMANE SOCIETY OF INDEPENDENCE COUNTY }
website / petfinder.com/shelters/AR65.html
facebook / @HSICPETS
location / Batesville

{ HUMANE SOCIETY OF THE OZARKS }
website / hsozarks.org
facebook / @hsozarkse
location / fayetteville

{ HUMANE SOCIETY OF PULASKI COUNTY }
website / warmhearts.org
facebook / HumaneSocietyPulaskiCounty
location / little rock

{ HUMANE SOCIETY OF SALINE COUNTY }
website / hssaline.org
facebook / @hssaline
location / bauxite

{ HUMANE SOCIETY OF SEARCY }
website / awos.petfinder.com/shelters/hss.html
facebook / @humanesocietyofsearcy
location / searcy

{ LAST CHANCE ARKANSAS }
website / lastchancearkansas.org
facebook / last chance arkansas
location / little rock

{ MAUMELLE FRIENDS OF THE ANIMALS }
website / maumellefoa.org
facebook / @MFOAFB
location / maumelle

\+

{ MORNING MEADOW ANIMAL RESCUE }
website / mmanimalrescue.weebly.com
facebook / morning meadow animal rescue
location / fordyce

\+

{ HEBER SPRINGS HUMANE SOCIETY }
website / awos.petfinder.com/shelters/AR56.html
facebook / heber springs humane society
location / heber springs

\+

{ NEEDY PAWS ANIMAL SHELTER }
website / petfinder.com/shelters/AR05.html
facebook / @needypawsanimalshelter
location / clarksville

\+

{ NO KILL RUSSELLVILLE }
website / nokillrussellville.org
facebook / @NKRussellville
location / russellville

\+

{ OUT OF THE WOODS ANIMAL RESCUE }
facebook / out of the woods animal rescue
location / fairfield bay

\+

{ OZARK HUMANE SOCIETY }
facebook / Ozark Humane Society
location / harrison

+

{ PETS ALIVE NORTHWEST ARKANSAS }

website / petsalivenorthwestarkansas.org
facebook / @Petsalivenwa
location / fayetteville

+

{ RESCUE ROAD }

website / rescueroad.org
facebook / @rescueroad
location / little rock

+

{ ROCK CITY RESCUE }

website / rockcityrescue.org
facebook / @rockcityrescue
location / north little rock

+

{ ROCKY RIDGE REFUGE }

website / rockyridgerefuge.com
facebook / @rockyridgerefuge
location / mountain home

+

{ SEBASTIAN COUNTY HUMANE }

website / sebastiancountyhumanesociety.org
facebook / @Sebastiancountyhumanesociety
location / fort smith

+

{ SPRINGDALE ANIMAL SERVICES }

website / petfinder.com/pet-search?shelter_id=AR91
facebook / @Springdaleanimalservices
location / springdale

\+

{ STOP ANIMAL CRUELTY IN HOT SPRING & GARLAND COUNTIES } - AR

website / sachsgc.com
facebook / Stop Animal Cruelty in Hot Spring & Garland Counties - Arkansas
location / hot springs

\+

{ WILES RESCUE ANIMAL SHELTER }

facebook / @wilesrescueshelter
location / highland

\+

{ ARKANSANS AGAINST PUPPY PILLS }

facebook / @ArkansansAgainstPuppyMills

\+

{ GAIL'S PETS SECOND CHANCE }

website / gailspets.org
facebook / Gails Pets Second Chance
location / oakland

\+

{ HUMANE SOCIETY OF CLARK COUNTY }

website / humanesocietyclarkcounty.org
facebook / humane Society of Clark County, Arkadelphia
location / arkadelphia

\+

{ STONE COUNTY HUMANE SOCIETY }

website / schspets.org
facebook / Stone County Humane Society
location / mountain view

{ THE LITTLE ROCK ANIMAL VILLAGE }
website / friendsoftheanimalvillage.org
facebook / @LittleRockAnimalVillage
location / little rock

{ ACKNOWLEDGEMENTS }

To my husband, Patrick: Thank you for adopting Jake, and for letting me be the crazy dog and cat lady that I am. You have always supported all the animals I have brought home as well as all my ideas—including the idea to write this book. You will remain my backbone and best friend. I love you.

Thank you to my entire family and friends for your love and support.

Thank you, Whitney Bower, for bringing the images in this book alive. Your talent is extraordinary and your love for dogs is unparalleled. Here's to all the hours, barking, and treat dangling we had to do in order for you to capture the perfect image!

Thank you to my wonderful editor, Erin Wood, for believing in this book and always knowing the correct word.

Thank you to my talented bonus sister, June Upton. You know how to make anything interesting and beautiful!

I am forever grateful for all the amazing people that submitted their dogs' stories. I wish I could have put every story in the book. Keep rescuing and supporting your local rescue organizations. We have to be the voices for the dogs who don't have them!

Lastly, Townsend Vest, you made me a mommy and I am forever grateful. Being your mom brings me a kind of joy I have never experienced before. Always rescue and help an animal in need. I love you so much.

{ CONTRIBUTORS }

GRACE VEST /

author

writer

+

mother

ERIN WOOD /

editor

editor & publisher

+

et alia

{ CONTRIBUTORS }

JUNE UPTON /

cover & layout - design

creative director

+

AROMATIQUE®

WHITNEY BOWER /

photography

photographer

+

WHITNEY BOWER

www.ingramcontent.com/pod-product-compliance
Lightning Source LLC
Chambersburg PA
CBHW042031150426
43199CB00003B/25